USA
NASA
Discovery

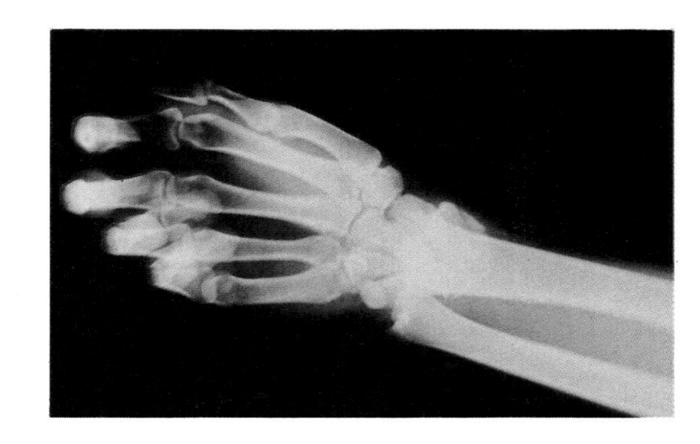

DISCOVER Light & Sound

Contributing Writer:
Francis Reddy

Consultant:
Isaac Abella

AN IMPRINT OF FOREST HOUSE™
School & Library Edition

Louis Weber, C.E.O.
Publications International, Ltd.
7373 North Cicero Avenue
Lincolnwood, Illinois 60646

Manufactured in U.S.A.

8 7 6 5 4 3 2 1

ISBN: 0-7853-0811-3

Photo Credits:

Front cover: **FPG International:** Gary Buss: (bottom); Jook Leung: (top); **H. Armstrong Roberts, Inc.** (center).

Archive Photos: 13 (bottom); Mary Bloom: 10 (top); **FPG International:** 38 (bottom), Back endsheet (left center); William D. Adams: 37 (bottom left); Jade Albert: 15 (left); Lee Balterman: 9 (top), 43 (left); Jose Luis Banus: 28 (left); Charles Benes: 18 (top); Tom Campbell: 18 (center); Color Box: Front endsheet (left center); Ed Cooper: 41 (top); Floyd Dean: 12 (top); Art Montes DeOca: 6 (bottom), 23 (bottom); Peter Gridley: 32 (bottom); Dennis Hallinan: 28 (top); Michael A. Keller: 9 (bottom); Michael Krasowitz: 20 (center); Lee Kuhn: 13 (center); Bill Losh: 36 (top); D.C. Lowe: 17 (left); Guy Marche: 20 (bottom), 22 (bottom), 43 (top); E. Nagele: 4 (center); NASA: 23 (top); Neil Nissing: 15 (bottom); Diane Padys: 21 (top); Maria Pape: 26 (top); John Pearson: 15 (right); Terry Qing: 37 (center); Ross Rappaport: 9 (center); Robert Reiff: 30 (bottom); Martin Rogers: Back endsheet (bottom); Ulf Sjostedt: 10 (bottom); C. Smith: 35 (top); J.E. Stevenson: 10 (center); Jeffrey Sylvester: 23 (center), Back endsheet (right center); Bob Taylor: 20 (top); Telegraph Colour Library: Title page, 7 (top & bottom), 8 (top & bottom), 12 (bottom), 14 (left), 18 (bottom), 21 (bottom), 28 (right), 39, Back endsheet (top left & top right); T. Tracy: Front endsheet (top left); U.S. Navy: 30 (top), 42 (bottom); Larry West: 34 (center); L. Willinger: 13 (top); Toyohiro Yamada: 32 (top); Jack Zehrt: Front endsheet (bottom); Nikolay Zurek: 22 (top); **International Stock Photos:** Wayne Aldridge: 31 (right); I. Wilson Baker: 26 (bottom); Jim Cambon: Table of contents (left center); Tom Carroll: 42 (center); Warren Faidley: 4 (bottom); A. Howarth: 36 (bottom); Peter Langone: 14 (top), 37 (top); Steve Lucas: 17 (right); Michael Philip Manheim: Front endsheet (top center), 5; Ronn Maratea: 4 (top); John Michael: Table of contents (bottom left), 37 (bottom right); NASA: 40 (top); Richard Pharaoh: 19, 31 (left); Phyllis Picardi: 16 (left), Back endsheet (top center); Lindsay Silverman: Table of contents (top right), 6 (top); Bill Stanton: 7 (center); Johnny Stockshooter: 11; White/Pite: 14 (right); **NASA:** 29 (left); **Rainbow:** Bob Curtis: 35 (bottom); Coco McCoy: 29 (bottom); Dan McCoy: Front endsheet (top right), 22 (center), 31 (top), 33 (top & center), 36 (center), 40 (center); Chris Rogers: 27; **Visuals Unlimited:** Table of contents (bottom right), 30 (center), 38 (center), 40 (bottom), 42 (top), 43 (right); Hank Andrews: 41 (right); Bill Beatty: 26 (center); A.J. Copley: 24 (left & right), 33 (bottom), 34 (bottom), 41 (left); John D. Cunningham: 29 (right), 34 (top); Carlyn Galati: 21 (center); Bill Kamin: 25 (left); Daphne Kinzler: 16 (right); Alan McClure: 38 (top); Glenn Oliver: 6 (center); SIU: Table of contents (right center); Doug Sizemore: Front endsheet (right center); Richard Walters: 25 (right).

Francis Reddy is a freelance science and technology journalist who has published five books and more than 70 newspaper and magazine articles for juvenile and general adult audiences. He has provided consultation, research, and writing for exhibits at the Museum of Discovery and Science in Ft. Lauderdale and the Atlanta Museum of Science and Technology.

Dr. Isaac Abella is Professor of Physics at the University of Chicago and is a member of the National Academy of Sciences Committee on Undergraduate Science Education and K-12 Science Standards. Dr. Abella has done research in laser physics and laser interactions in matter, and won the Quantrell Prize for Excellence in Undergraduate Teaching.

Illustrations: Peg Gerrity; Lorie Robare.

CONTENTS

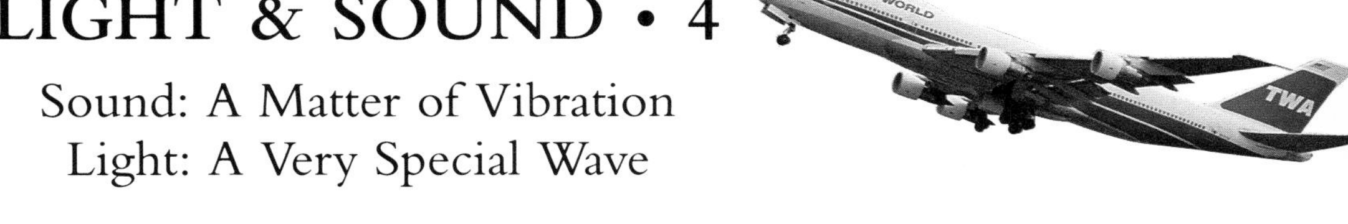

LIGHT AND SOUND

bring us our world. Our eyes and ears tell us most of what we know about the things around us. These organs are sensitive to the waves we call light and sound.

Light waves bring us the flicker of a flame and the beauty of a sunset. Sound waves carry the voices of friends and the tune of a favorite song.

But what our eyes can see and our ears can hear is only a small part of the waves that swirl all around us. There are waves beyond our senses. We can communicate almost instantly through radio waves, map storm clouds using heat, and explore the universe with X rays. We can use sound to help understand our world. Let's explore the world of light and sound!

SOUND: A MATTER OF VIBRATION

The sound of an airplane taking off is enough to make you run for cover, while a bird's song might make you want to move closer and listen. Both sounds are produced by waves traveling through the air.

All the sound we hear, from the rustling of leaves to the roar of a jet engine, is caused by vibrations in some substance.

These rippling waves are transporting energy that will eventually affect areas very far away from the original disturbance.

Sounds are waves that travel through the air, just as ripples travel across a quiet pond after a stone is thrown into the water.

Although they behave in similar ways, sound waves moving in air are different from water waves. Water waves are called *transverse* (trans-VERS) waves. They cause the water to move up and down. You can make transverse waves by tying down one end of a rope and giving the other end a quick jerk up and down. As the wave pulse travels along the rope, it moves up and down with the wave's movement.

Sound waves do not travel this way through air. When a sound wave travels through a substance, the particles that make up the substance are briefly squeezed together and then spread apart. Places where the particles are more crowded than they would be without the wave are called *compressions* (kuhm-PRE-shuhns). The areas where particles are more spread out are called *rarefactions* (rar-uh-FAK-shuhns). This is called a *longitudinal* (lawn-juh-TOOD-nuhl) wave because the particles move back and forth in the same direction as the wave's motion. To picture this type of wave, imagine giving a quick push-pull to the end of a spring. A tight pulse of coils moves down the spring (compression) followed by a few coils spaced farther apart than those along the rest of the spring (rarefaction).

Sound can move through any substance, whether it's a solid, liquid, or gas. It travels fastest through the materials that most resist being compressed. This means that sound has the highest speed in solids. It moves more slowly in liquids and slowest of all through gases such as air.

The actual speed of sound in any location depends on the temperature, pressure, and humidity of the air. Sound travels much more slowly than light.

What effect does this speed difference have on the way we perceive things? Across the distance of a soccer field, for instance, it means there is a moment's delay between seeing something happen—such as a player kicking the ball—and hearing the sound it produces. This delay gets longer as distances increase. Although light reaches us almost instantly, sound takes about five seconds to travel one mile. This delay can be used to figure out the distance to the sound's source.

The branch of science that deals with the study of sound is called *acoustics* (uh-KOO-stiks). The study of acoustics affects our lives in many ways. It's important in understanding how people speak and hear. The behavior of sound waves affects the performance of music and the construction of speakers and recording equipment. Companies that build vehicles use acoustics to reduce engine noise. The study of sound is everywhere!

In most weather conditions, sound waves travel through the air at speeds of about 760 miles per hour.

Sound cannot travel through space because there is no material for the waves to squeeze together.

Acoustics is very important in the design of theaters.

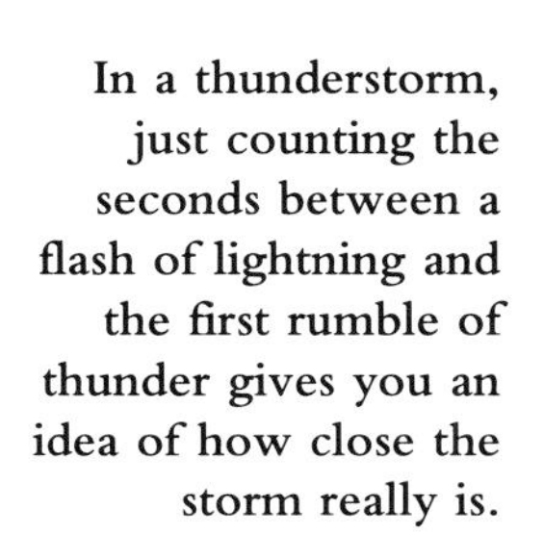

In a thunderstorm, just counting the seconds between a flash of lightning and the first rumble of thunder gives you an idea of how close the storm really is.

LIGHT: A VERY SPECIAL WAVE

No spacecraft today can even come near the speed of light.

Like sound, light is a form of energy that travels in waves. It's a special kind of energy called *electromagnetic* (i-lek-tro-mag-NE-tik) energy. But light waves do not move by shifting matter back and forth. Light waves are electric and magnetic waves that actually move most quickly where there is no matter. Our eyes can detect only a small part of the electromagnetic waves that surround us. Radio waves, heat, and X rays are all types of electromagnetic waves invisible to our eyes.

Light is the fastest thing in the universe. It travels about 186,000 miles per second! The sunlight we see has been traveling for about eight minutes, and light from the nearest star takes over four years to reach our eyes.

Light from the moon takes about 1.3 seconds to reach us.

Light energy can come from different sources. Most often, light comes from energy changes within individual atoms.

Each atom consists of a center, called a *nucleus* (NOO-klee-uhs), and particles called *electrons* (i-LEK-trons) that orbit the nucleus. Some electrons have orbits close to the nucleus, and others have larger orbits that take them farther from the nucleus. When an electron gains energy, it jumps out to a higher level or orbit. The electron must lose energy to fall back to a lower orbit. When an electron loses energy, it emits a small amount of light energy called a *photon* (FO-ton).

When astronomers view a distant galaxy, they are seeing light that has been traveling many *millions* of years.

A common source of light is the glow from a heated piece of metal. As the temperature of the metal rises, the atoms and molecules within it move faster and faster. They bump into one another with greater speed. Each collision gives energy to the electrons and forces them into higher orbits around their atoms. When they fall back, they send out photons of light. At first, this light is invisible to us. Then, as the metal gets hotter, electrons absorb greater amounts of energy, leap to higher orbits, and give off more energy.

As a metal is heated, it will look red, then orange, yellow, and finally yellow-white.

A lightbulb contains a piece of glowing metal. In this case, the metal is a thin wire called a *filament* (FI-luh-muhnt). When the lamp's switch is turned on, an electrical current runs through the wire. The current heats the wire and jiggles the electrons in the wire. The electrons gain energy and move into higher orbits. When they fall back down to their original levels, the electrons give off light photons.

Many living things, such as fireflies, produce their own light through chemical reactions.

Both the lightbulb and the heated metal are examples of *incandescence* (in-kan-DE-suhns), light given off by materials brought to a very high temperature.

When electricity flows through the tubes, the neon and other gases in this sign give off light.

Some materials can be made to give off light without being intensely heated. This process, called *luminescence* (loo-muh-NE-suhns), comes from atoms and molecules that have gained energy without much heating and that release this energy as light.

Many lightbulbs have filaments made of a metal called tungsten.

SOUNDS AROUND YOU

have an effect on how you behave. You seek out some sounds and try to avoid others. People listen to music because it pleases, relaxes, and entertains them.

But many of the sounds we hear every day may have just the opposite effect. The roar of a jet airplane. Blaring sirens and honking horns in city traffic. Many people feel uneasy if forced to listen to these noises for long.

The sounds we think of as musical tones occur in simple and regular patterns. These sounds include tones made by instruments such as the trumpet or piano. Sounds we think of as being noise include the crash of cymbals and even spoken words. These sounds are more complex than musical tones.

LOOKING INTO SOUND

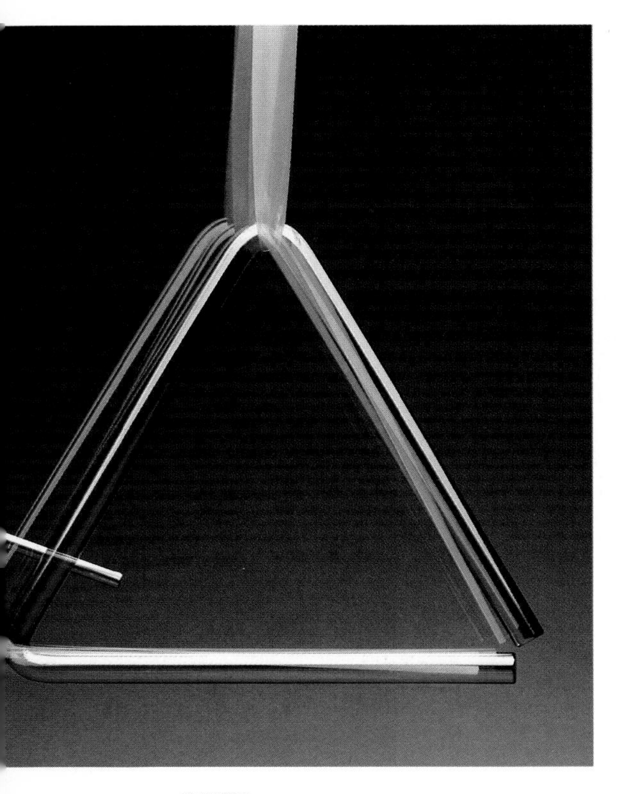

We picture sound waves by looking at the vibrations that make them.

If you strike a musical triangle, you can see the vibrations that produce sound waves in the air.

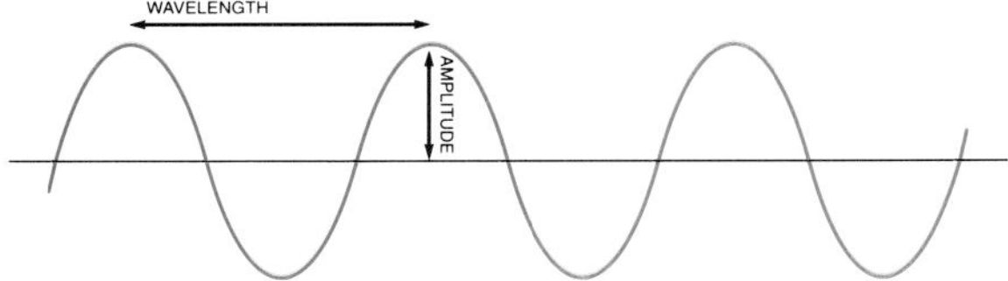

The sound of the space shuttle taking off is so loud it could damage your ears. The takeoff produces sound waves with a huge amplitude.

We know that sound travels through a substance as a series of compressions of that substance. This type of wave is difficult to imagine. But you can picture a way to draw a sound wave.

Many musicians keep their instruments in tune by checking them against the standard tone produced by a *tuning fork*. This small, U-shaped metal bar produces a pure tone when made to vibrate by a gentle tap. Its vibrating prongs set the air into motion, creating the sound we hear. Imagine that we attach a pen to one of the prongs of a tuning fork. While the fork vibrates, we move the pen straight across a piece of paper. The tuning fork draws a wavy line that charts its own vibrations. It's a picture of the back-and-forth movement that makes the sound waves we hear.

Let's look at this line more closely. The height above or below the center of each wave is called the wave's *amplitude* (AM-pluh-tood). At first, the tuning fork's prongs are moving back and forth a great distance; the waves have a large amplitude. The sound is also at its loudest. As the vibration fades, the amplitude of the sound waves decreases and we hear a softer tone.

The distance between the tops of the waves is exactly the same even though the amplitude gets smaller. This distance is called the *wavelength* of the sound.

Another aspect of sound waves is called *frequency* (FREE-kwuhn-see). Frequency is measured as the number of complete wave cycles that pass by a single point every second. A sound with a frequency of one wave cycle per second has a frequency of 1 hertz (Hz). The lowest tone on a piano keyboard has a frequency of 27 Hz, while the highest vibrates at 4,200 Hz.

A sound's frequency and wavelength are related. A sound with a long wavelength has a low frequency and a low pitch. A sound with a short wavelength has a high frequency and pitch. It's natural to arrange the sounds the human ear can detect in the same way as a piano keyboard is set up—from the lowest frequency to the highest.

The range of frequencies we can hear is quite impressive, but many animals can hear sounds we cannot. Frequencies higher than those we can hear are called *ultrasound*. Frequencies lower than the limit of human hearing are called *infrasound*.

Like all waves, sound waves can be reflected and absorbed by the obstacles they run into. People who design theaters and concert halls must pay careful attention to the way sounds bounce around. For example, the ground and audience at outdoor concerts absorb so much sound that bands and orchestras often play inside a shell-shaped structure that reflects sound back to the concert-goers.

Elephants make sounds of far lower frequency than we can detect, although no one knows exactly why.

Dogs can hear frequencies higher than 30,000 Hz.

Human ears can pick up frequencies between 20 Hz and 20,000 Hz.

Even a band shell isn't always enough for a large crowd, and today microphones and speakers are usually used to amplify the sound.

MUSIC MAKERS

Membranophones ("skin sound") are instruments played by striking a skinlike covering stretched over the opening of a hollow object. Most membranophones are drums.

Musical instruments are often classified based on how their vibrations are produced.

All stringed instruments are classified as *chordophones,* a word that means "string sound." Chordophones include the violin, harp, and piano.

Aerophones ("air sound") are wind instruments such as the saxophone. Their sound comes from a vibrating column of air set into motion when someone blows into the instrument.

Musical instruments range from the human voice to electronic synthesizers. Noise is made up of a jumbled mixture of sound frequencies. Musical notes, however, consist of one strong fundamental frequency plus other frequencies that give a unique "color" to the sound of each instrument. For example, no one would confuse a violin with a trumpet, even when both play the same note.

Most musical instruments have two things in common—a vibrating source that actually produces the sound and an object called a *resonator* (RE-zuhn-ay-ter) that helps amplify and color the sound.

The sounds from the vibrating source in many instruments are actually not very strong. The sounds are made louder by *resonance* (RE-zuhn-uhns). Resonance is a way of getting a large vibration with a small effort.

All objects tend to vibrate at certain natural frequencies. An object's natural frequency is also called its *resonant frequency.* If an object can be made to vibrate at its resonant frequency, the amplitude of its vibrations increases dramatically with only a little additional effort. Anyone who has tried out a playground swing has used resonance. To get a swing to go as high as it can, you push it at the right time and in the right direction to match its resonant frequency.

Most musical instruments amplify their sounds by the same principle. *Resonators* are often large solid or hollow objects attached to the source of the vibration in a musical instrument. They amplify and prolong the sound from an instrument. They also add distinctive tones by emphasizing some frequencies over others. This gives the sound from each instrument a unique color.

The main resonator for all wind instruments is the column of air within them. The air within the hollow body of a guitar or violin has one resonant frequency, while the body itself has another. Both are important in amplifying the very faint sound of vibrating strings.

The sound of the human voice comes from an organ called the *larynx* (LAR-inks), or "voice box," which is located in the tube that connects the lungs to the throat. The largest part of the larynx forms the "Adam's apple," a ridge on the neck just above the collarbone. You can feel vibrations from the larynx by lightly touching this bump as you speak.

The human voice benefits from resonance, too. The throat, mouth, nasal sinuses, and chest all influence the final sound. Each of these cavities has a resonant frequency that depends on two things—the volume of air it contains and the size of any openings to the outside. Only the mouth can change both of these properties.

The human voice is an instrument almost anyone can use for making music.

The human voice is a unique and very flexible music maker.

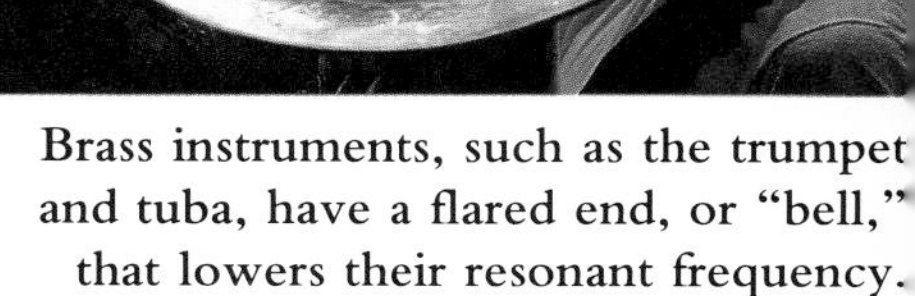

Brass instruments, such as the trumpet and tuba, have a flared end, or "bell," that lowers their resonant frequency.

You can enjoy a guitar's music because resonance amplifies its sound. Without resonance, the sound of the strings would be too faint to hear from more than a few inches away.

EARWORKS

The Ear

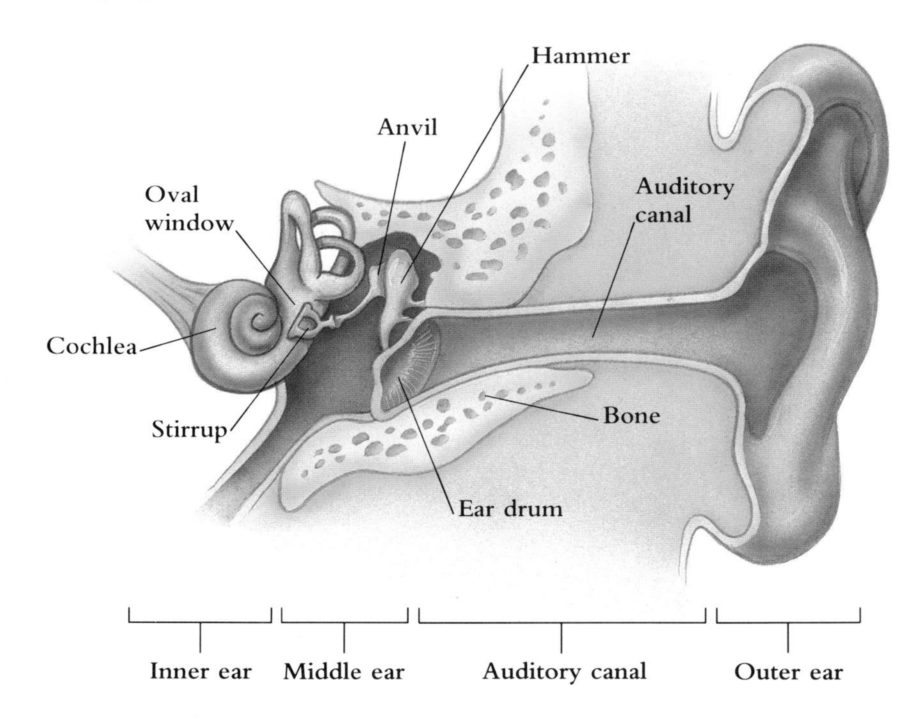

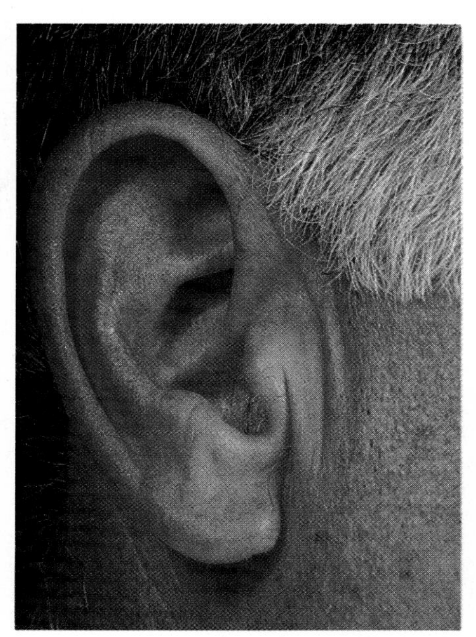

Only mammals have an outer ear.

Your ears detect the vibrations in the air that your brain translates into sounds.

The ear translates the tiny vibrations of the air into electrical signals that are sent to the brain. We hear the amplitude of sound waves as loudness: The greater the amplitude, the louder the sound. We hear the frequency of sounds as *pitch,* the highness or lowness of a tone. The higher the frequency, the higher the pitch. And we hear particular mixtures of frequencies as tone. This quality helps us tell the sounds of musical instruments apart, even when they play the same note.

There is much more to the ear than meets the eye, but the part we do see is called the outer ear. The size and shape of the outer ear help collect and funnel sound waves into the opening of the ear. This opening leads into the *auditory canal*, a tube about one inch long. The other end of the auditory canal is sealed off by the *eardrum,* a thin membrane that vibrates as sound waves strike it.

On the other side of the eardrum, in what is called the middle ear, lies a chain of three tiny bones. Nicknamed the "hammer," the "anvil," and the "stirrup" because of their shapes, the bones rattle whenever the eardrum vibrates. They carry sound vibrations to the inner ear. The last of the bones, the stirrup, rests against a membrane-covered opening, called the *oval window,* in the wall of the inner ear.

The inner ear consists of a snail-shaped, fluid-filled organ called the *cochlea* (KOHK-lee-uh). This organ translates waves of pressure into nerve impulses to the brain.

The *audio spectrum*—the range of frequencies the ear can detect—lies between 20 Hz and 20,000 Hz, although most people cannot hear a pure tone at either extreme. The ear is not very sensitive to sounds lower than 100 Hz. This probably helps us ignore sounds produced in our bodies, such as heartbeats and blood rushing through veins. The ear is very sensitive to sounds near the resonance frequency of the auditory canal, between 3,000 Hz and 4,000 Hz.

One useful scale scientists and engineers use to measure sound is called the *decibel* (DE-suh-bel). Since the ear actually hears sound by detecting changes in pressure, the decibel scale measures the amount of pressure sounds put on the ear. The softest sound the human ear can detect is 0 decibels. You begin to feel pain in your ears when you hear a sound at about 130 decibels. In between, the average background noise in a home is about 40 to 50 decibels, heavy street traffic is about 80 decibels, and a rock concert is about 120 decibels. A jet engine running at full power is a painful 160 decibels.

The sounds received by each ear are not exactly the same. Some sounds may arrive at one ear before the other. High-frequency sounds seem softer to the ear farthest from the source. The brain combines the sounds heard from both ears and uses these differences as clues to tell us what direction the sounds came from.

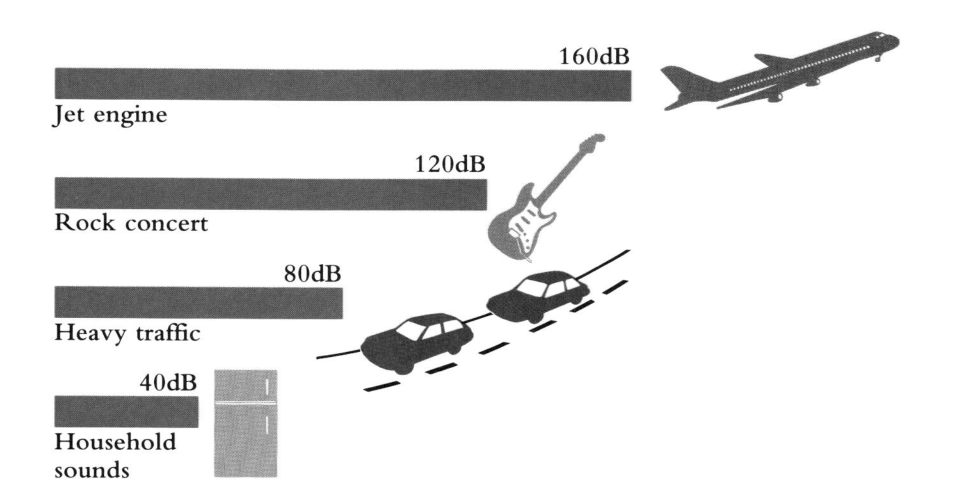

A "loudness spectrum" illustrates how much sound energy it takes to make the human ear take notice.

Sound travels much faster in water, so scuba divers hear none of the differences that help us tell where a sound comes from.

A jet engine at full power can rupture eardrums.

A WORLD OF LIGHT surrounds us. If you shine a ray of what we think of as "white light" through a prism or a raindrop, the light will separate into a band of colors that range from red to violet. Like sound waves, light waves have wavelength and frequency.

Red light has the lowest frequency and the longest wavelength. Violet light has the highest frequency and the shortest wavelength of visible light. Between these wavelengths lies all the light our eyes can detect.

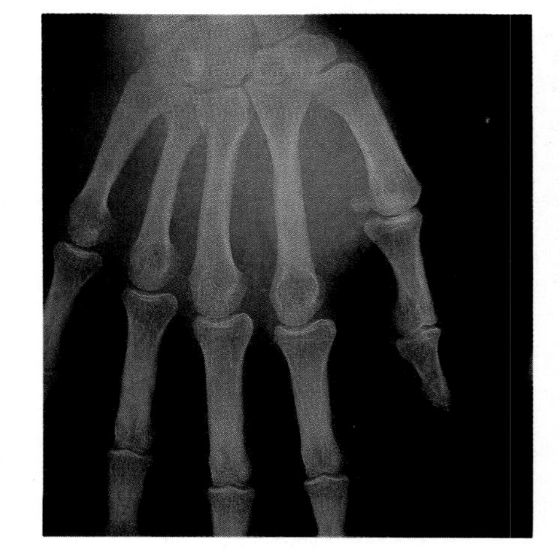

But there are many kinds of electromagnetic waves our eyes cannot see. Think of the X rays doctors use to see through teeth and bone. The waves we feel as heat, radio waves, and the microwaves we use to cook our food are also electromagnetic waves we cannot see.

20

MORE THAN LIGHT

We are surrounded by many kinds of electromagnetic radiation that our eyes cannot detect.

Natural electricity, in the form of lightning, can make the unpleasant bursts of radio noise we call static.

What we call "visible light" is actually part of a more general energy system called *electromagnetic radiation.* In fact, visible light is only a small part of the *electromagnetic spectrum,* the whole range of electromagnetic radiation. Visible light and waves in other parts of the spectrum are different in three ways: They have different frequencies, different wavelengths, and photons with different amounts of energy. Waves of higher frequency, such as X rays, have shorter wavelengths and photons with higher energy. Lower frequency waves, such as radio waves, have long wavelengths and weak photon energy.

Radio waves are the longest of the spectrum. Radio and television broadcasts aren't the only use for radio waves. Very short radio waves, called *microwaves,* can be used to measure distances to faraway objects. A transmitter sends out a pulse of microwaves. Some of them may bounce off objects such as airplanes, and then return to their source. Because the waves all travel at the speed of light, measuring the time between the pulse and the returned signal will give the object's distance. This method is called *radar.*

Many people cook their meals with microwaves. In a microwave oven, radio waves about 4.7 inches long have different effects on water molecules in the food and other molecules in the food containers. The food is heated but the containers stay cool.

In a microwave oven, the water molecules in the food absorb energy and heat up. But a plastic or ceramic dish that contains the food is not affected.

Radar antennas send out microwaves to measure distances to far-off objects.

Waves that are much shorter than microwaves, but still longer than red light, are called *infrared*. We feel infrared waves as heat. About 60 percent of the sun's electromagnetic waves occur in the infrared part of the spectrum.

Ultraviolet waves have wavelengths that are shorter than violet light. Ultraviolet wavelengths carry a great deal of energy, enough to break apart many of the molecules found high in the Earth's atmosphere. Most of the sun's ultraviolet light is absorbed in this way by the ozone in the atmosphere.

X rays are even more energetic and actually pass through many materials. They are used in medicine to see inside the body. These waves are also given off by the hottest gases in the sun and stars, but none reach Earth's surface because the atmosphere absorbs them. X rays are more dangerous than ultraviolet light because they can more easily break up molecules and damage cells in the body.

Gamma rays are the most powerful and penetrating electromagnetic waves. Gamma rays are given off only when the nucleus of an atom changes its energy. Gamma rays are very damaging to living tissue. Diseased cells are more sensitive to them than healthy cells, so gamma rays are sometimes used to treat cancer. Although our atmosphere screens out gamma rays from space, satellites have detected them from exploding stars and distant galaxies.

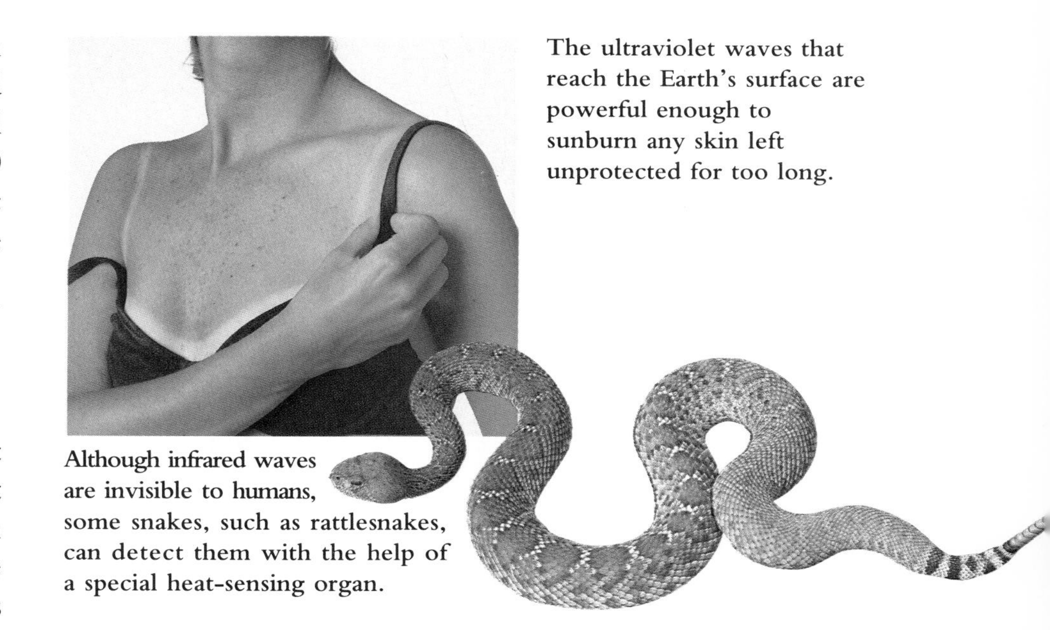

The ultraviolet waves that reach the Earth's surface are powerful enough to sunburn any skin left unprotected for too long.

Although infrared waves are invisible to humans, some snakes, such as rattlesnakes, can detect them with the help of a special heat-sensing organ.

Most incandescent materials—such as a lightbulb or lava from a volcano—actually give off more heat than light.

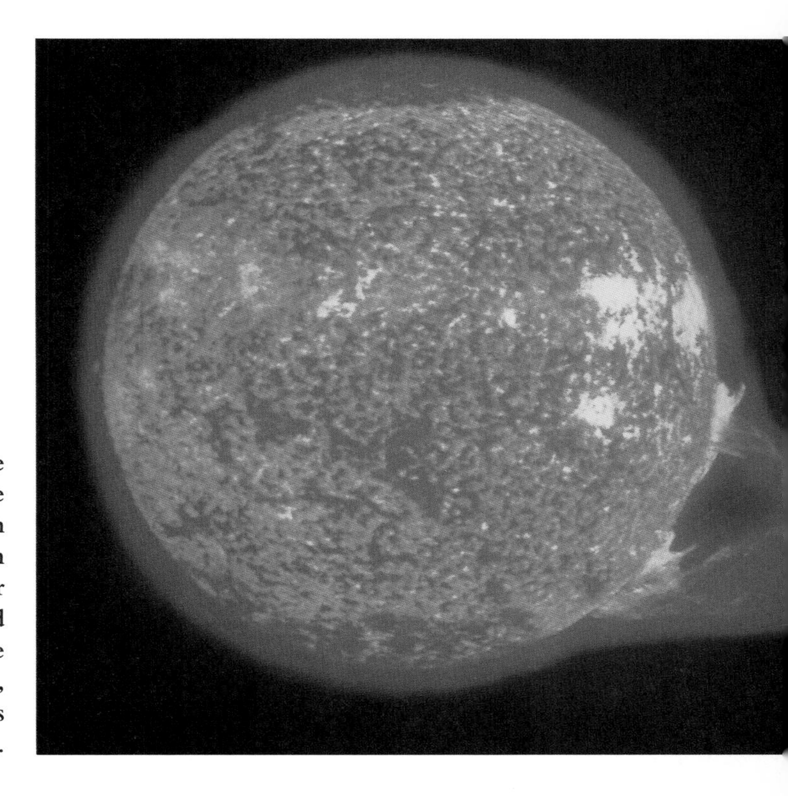

Gamma rays are given off in the cores of stars, such as our sun, in nuclear power plants, and in particle accelerators, powerful devices that smash atoms.

USING LIGHT

Right: Photos are often processed in a special room called a darkroom. Below: This image is called a *negative* because it is the opposite of the scene that was photographed—the dark areas look light and the light areas look dark.

We know that photons are the basic bits of light energy. X-ray and gamma-ray photons carry enough energy to break up molecules or alter the chemicals that make up living cells. Visible-light photons usually aren't powerful enough to make such changes, but there are several important exceptions.

One of the most familiar is photography. Photographic film is coated with one or more layers of light-sensitive chemicals. When a camera takes a photograph, the photons absorbed by these chemicals produce invisible changes in them. When film is processed, the portion of film exposed to the brightest light turns darkest, and the portion exposed to the least amount of light remains clear. It takes another process using light-sensitive paper to make the pictures we pick up at the photo store.

Green plants found a way to use light energy billions of years ago.

Green plants use light energy to drive the chemical cycles that make their food. They absorb the energy in sunlight and then use it to power chemical reactions that turn raw materials into food for the plant. This process is called *photosynthesis* (fo-to-SIN-thuh-suhs).

Green plants contain the pigment *chlorophyll* (KLOR-uh-fil), a light-sensitive chemical. This chemical is used in photosynthesis.

In one way or another, most of the life on Earth depends on photosynthesis. From water and carbon dioxide in the air, plants make a kind of sugar. It provides food not only for the plant but also for any animals that eat the plant. Plants release oxygen, the other product of photosynthesis, into the atmosphere. Animals breathe in oxygen to process the foods they eat.

Our modern machines rely on the sunlight captured by ancient plants. Hundreds of millions of years ago, great forests grew all over the world. As plants lived and died, thick blankets of leaves, branches, and other plant matter were buried on the forest floor. Over time, this matter was compressed by the newer layers of plant material and by mud and rock that settled on top of it. This process continued for hundreds of millions of years, and today much of this plant matter has been chemically transformed into coal. Similar processes formed oil and natural gas from ocean plants. Oil, gas, and coal are called fossil fuels because of their connection to ancient life. Most of our energy comes from the burning of fossil fuels.

Devices called solar cells can convert light directly into electricity. They work because some materials, like silicon, can absorb incoming photons of light. This causes the silicon's electrons to change in such a way that they can conduct electricity.

How practical is solar energy as an energy resource on Earth? There are some problems. The efficiency of devices that convert solar energy into electricity is only about 15 percent. This means that it takes many solar panels to produce useful amounts of electricity. But scientists are studying ways of improving methods of using solar energy. The sun seems like a good bet for a practical, pollution-free energy source for the future.

Machines as different as a calculator and NASA's Hubble Space Telescope are powered by solar energy.

All the energy in fossil fuels comes from sunlight captured by plants in the forests of the Earth's past.

Large areas have to be covered with solar collectors to capture enough sunlight to generate much electricity.

SEEING CLEARLY

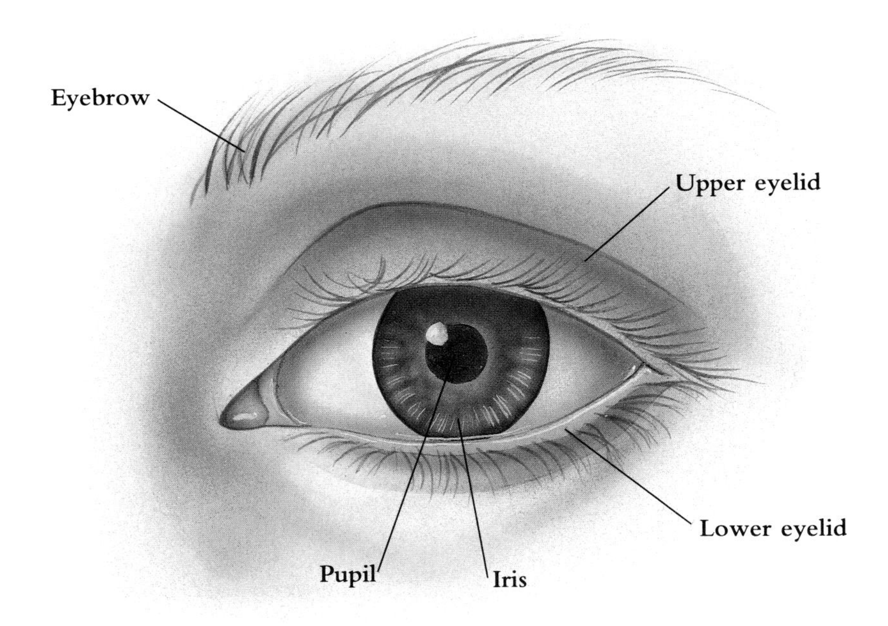

Vision is the sense we depend on most in our day-to-day lives.

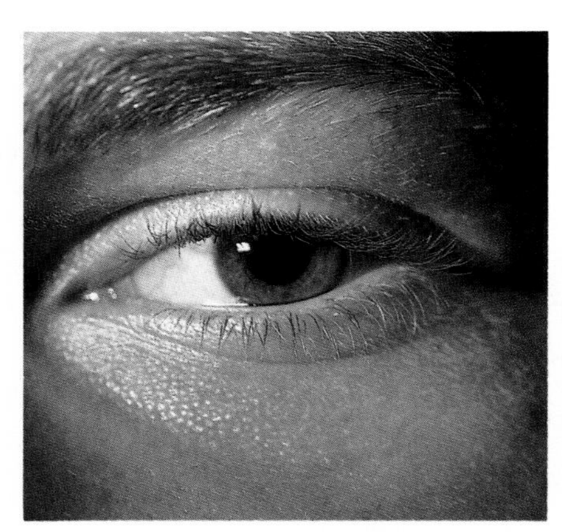

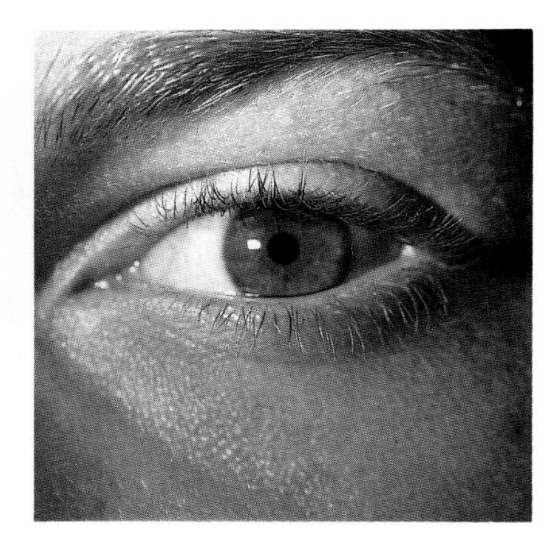

The iris expands or contracts in response to changes in the brightness of light.

The eye translates the vibrations of visible light waves into electrical signals that are then sent to the brain. We can see the wavelengths of light because light-sensitive chemicals line the inside surface of the eye.

A clear, dome-shaped layer called the *cornea* (KOR-nee-uh) first bends light waves as they enter the eye. It works like the lens of a camera, bending light waves so that they will come together as a focused image at the back of the eye.

Not all of the light passing through the cornea actually enters the eye. Behind the cornea is a colored ring of muscles called the *iris* (I-ruhs). When people mention the color of your eyes, they are really talking about the color of your irises. At the center of the iris is a hole called the *pupil.* Light actually enters the eye through the pupil. The iris changes the size of the pupil to help the eye adjust to changes in brightness.

After entering the pupil, light waves next reach the *lens* of the eye. Although the cornea does most of the work to focus incoming light, it needs some help bending the light from objects closer than about 20 feet. Tough muscles pull on and flatten the lens when the eye concentrates on a distant object. When the eye looks at something closer than about 20 feet, the muscles reduce their pull on the lens. As the muscles ease up, the lens becomes slightly thicker and better able to bend light.

Once through the lens, light passes through a jellylike fluid that fills the interior of the eyeball. Light penetrating even this far into the eye still has not been "noticed"—the eye hasn't *seen* anything. Light must strike a layer of cells at the back of the eye before it is actually detected. This layer is called the *retina* (RE-tuhn-uh).

The wafer-thin retina is made up of about 130 million light-sensitive cells, called rods and cones, plus other cells that send nerve impulses. Cones can see colors, but they work well only in bright light. Rods are very sensitive to weak light and outnumber cones 17 to 1. But rods cannot tell colors apart. Night scenes seem colorless to us because the light entering our eyes is too faint to stimulate the cones. In some animals, such as dogs, the layer behind the retina is very shiny. Light striking this layer is bounced back through the retina's cells to give the dog's light-sensitive cells a second chance of seeing it. This gives dogs and other animals very good vision in darkness.

Rods and cones contain chemicals called *pigments* that undergo changes when light strikes them. The stimulation of these pigments throughout the retina results in electrical signals that are flashed to the brain. Different parts of the brain deal with different features of the images we see, such as color, movement, and shapes. Each eye sees a slightly different image, which the brain merges into a single three-dimensional view.

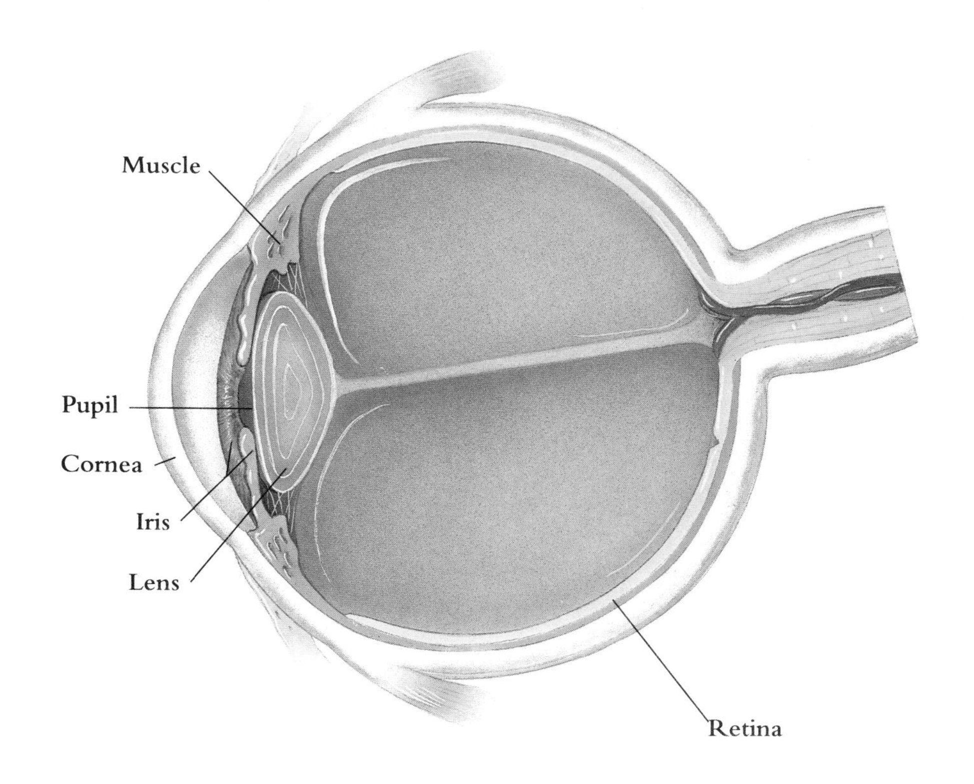

Possibly as much as two-thirds of the activity in the brain is involved in studying information that comes from the eyes.

Light bouncing back through the retina makes the eerie "eye glow" you sometimes see from animals at night.

Insects have *compound eyes*, honeycomb-shaped clusters of lenses. This type of eye is very sensitive to movement, but no one knows what kind of image the insect sees.

WAVES OF SOUND AND LIGHT

can do many things. The waves can be reflected by obstacles and sent back toward their source. Every glance in the bathroom mirror is a glimpse at the reflection of light. Waves can change their paths as they move from one substance into another. Lenses in a camera and the human eye bend light to place a clear image where it's needed. Other properties of waves let us use lasers to create three-dimensional pictures and keep instruments in tune.

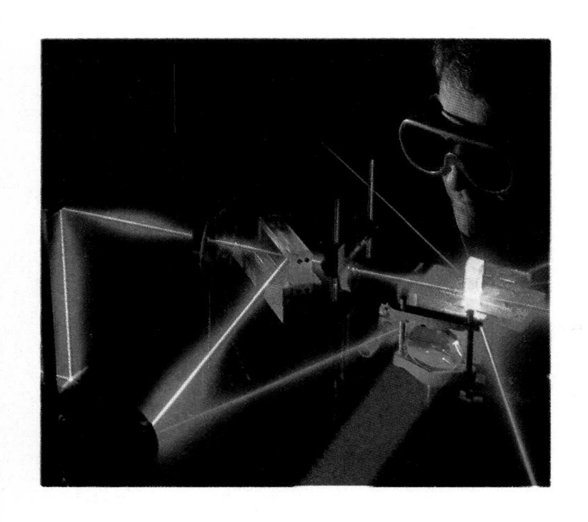

The properties of sound and light waves touch our lives in so many ways. Understanding how waves work helps us learn about and change our world.

REFLECTION

If everything reflected all the wavelengths of light equally, then the world would lack color and we would see only shades of gray.

Clouds, fresh snow, and the paper on this page look white because they reflect nearly all the light of every color that falls on them.

Whenever light encounters the surface of any substance, some of the light is absorbed, some goes through if the substance is transparent, and some bounces off. The last of these is called *reflection*. We see objects because our eyes detect the light that reflects off them. Transparent substances, such as air and clear glass, reflect and absorb little light.

In order for a substance to look colored, it must have a different effect on some of the wavelengths of light. Grass absorbs red and blue light, so the only color left to be reflected is green. Flowers, dyes, and paints absorb some wavelengths and reflect the ones that make up the colors we see.

Metals are very good reflectors. The first mirrors were simply polished pieces of metal. The familiar bathroom mirror is a combination of metal and glass—a thin film of reflective metal coats the back of a glass plate.

Light reflected from a very smooth surface goes in a definite direction that depends on the direction of the original waves. Light striking the shiny surface of a mirror bounces off it at an angle equal to the angle at which it hits the surface. The reflected light behaves as though it were coming from a point behind the mirror and traveling through a clear window. Your image in a flat mirror always seems to be as far behind the mirror as you are in front of it.

Each paint reflects different wavelengths of light. We see the colors of light that are reflected by the paints.

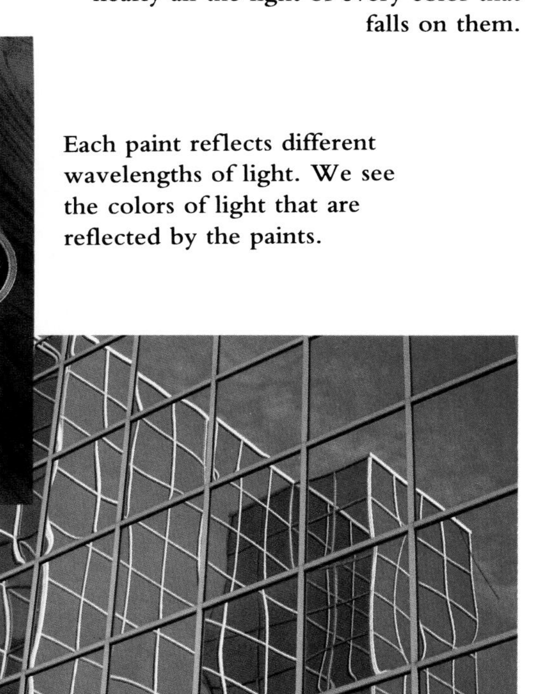

This building's windows are made reflective by a special coating on the glass. But the windows are not mirrors. If you stand close enough, you can see inside the building.

Waves can be concentrated or spread out by curved reflectors. A *concave,* or bowl-shaped, mirror causes the light reflected to come together at a focus. "Magnifying mirrors," or the bowl of a spoon, are concave. Concave shapes do the same for sound waves, too. For example, satellite dishes are concave and concentrate radio waves to a focus.

Dome-shaped, or *convex,* mirrors spread out their reflected light. The image reflected in the back of a spoon looks smaller—as though it were very far away from the reflector. For this reason, the convex rear-view mirrors that come with many cars have a warning written on them: "Objects in mirror are closer than they appear." The weird images made by fun-house mirrors come from changing the shape of the mirror surface.

Reflections are useful to some animals. Dolphins, insect-eating bats, and other animals use sound reflections in a way that's similar to radar. They send out pulses of ultrasound at frequencies higher than 50,000 Hz. The waves reflect off obstacles and carry information about the animal's surroundings, especially the location of food.

Sound waves striking a smooth wall reflect in much the same way. If you shout at a wall, say in a tunnel, the sound reflects back to you. The sound returns to you delayed by exactly the amount of time it would take if it really had been made by someone next to the tunnel wall. This is an echo.

Reflected radio waves have been used to make images of the hidden surface of the planet Venus. Light cannot penetrate the thick clouds that cover Venus, but the clouds do not interfere with radio waves.

Bats use echoes to locate objects.

We use radar to detect distant objects by sending out pulses of radio waves and sensing the waves reflected back.

If you stand some distance from a cliff and shout, the sound reflects off the cliff just as though someone on the other side had shouted to you at the same time.

REFRACTION

Right: People far from an explosion may hear it clearly while people nearby may not hear it at all. Below: The light waves that reach our eyes from the pencil change directions when they cross the water, the glass, and then the air.

All waves can experience refraction, but the most familiar examples are those involving light.

Waves traveling through one substance may change their path when they pass into another substance. Whether or not the waves bend from their original path depends on the speed of the waves in each substance. If the waves traveling through the first substance travel faster or slower through the second substance, then the direction of the waves will change. This is called *refraction* (ri-FRAK-shuhn).

The speed of sound in air depends partly on the air's temperature. Sound waves from a loud source, such as an explosion or thunder, travel at different speeds when they encounter air at different temperatures. Under the right conditions, waves traveling up and away from the ground can be refracted so that they head back down toward the ground. A sound could be more easily heard farther from its source than close by.

Light can also be affected by changes in the air. Changes in air density can cause light to bend. Air is thickest at the bottom of the atmosphere and becomes thinner higher up. When we look at a sunrise or sunset, we're seeing light that has traveled through many miles of the thickest air. The atmosphere bends light from the top edge of the sun so it looks higher than it really is—but the thicker air closer to the ground bends light from the sun's bottom edge even more. The result is a sun that looks not round but oval!

When the sun hovers at the edge of the horizon just before sunset, we can see it because of refraction. If the light from the sun were traveling in a straight line, the sun already would have disappeared from sight.

Refraction also helps us see more clearly. The shapes of different lenses affect the way light passes through them. Concave lenses, which are thicker on the ends than in the middle, spread out light waves. People who are nearsighted use concave lenses to help them see clearly. Convex lenses, which are thicker in the middle than at the ends, bring light rays together. People who are farsighted use convex lenses to correct their vision.

Convex lenses can be used as magnifying glasses.

All the wavelengths in visible light are not refracted equally. Red light is bent the least, and violet light is bent the most. This lets us use refraction to separate the colors of "white" light. If you shine a beam of white light through a wedge-shaped piece of glass, or *prism,* the colors of the rainbow fan out from it. The prism shape refracts each color of light by a different amount.

In order for you to see a rainbow, a curtain of water drops must be in front of you and the sun must be behind you.

Rainbows themselves are created by a combination of refraction and reflection. Sunlight striking a drop of water refracts as it enters, then reflects off the inside of drop, and finally refracts again as it leaves. The two refractions separate the colors of sunlight, and the reflection sends the light back toward the sun—and us.

Telescopes look outward into space. They use either a large lens (refracting telescope) or a large concave mirror (reflecting telescope) to collect light. The largest refracting telescope has a lens that is 40 inches across.

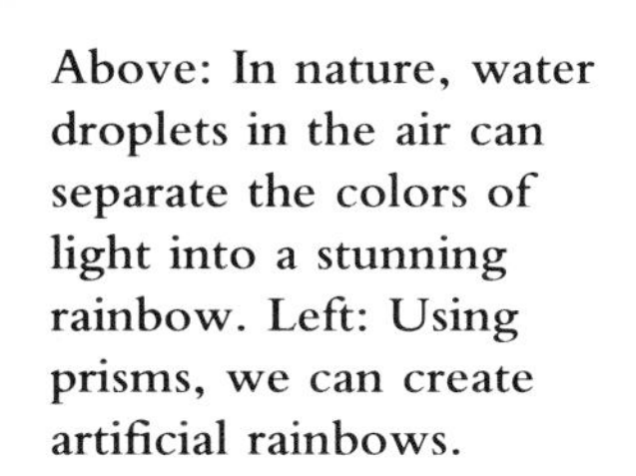

Above: In nature, water droplets in the air can separate the colors of light into a stunning rainbow. Left: Using prisms, we can create artificial rainbows.

INTERFERENCE

For two waves to be canceled out completely, both must have the same frequency and amplitude.

Water waves may combine to make a calm sea.

What happens when waves combine? It's possible for sound waves from different sources to combine and make silence.

Suppose two identical stones are thrown into a calm pond at the same time. If the two waves meet as each reaches the high point of its vibration (the crest of the wave), they will join together to make a double-high wave. But if they meet in the opposite way, with one wave at its crest and the other at its low point, they simply will cancel each other out. The low point of one wave is just as deep as the crest of the other is high. So when they are added together, the result is no wave.

The meeting of waves is called *interference,* and it happens in all waves. *Constructive interference* occurs when waves meet in such a way that they combine to make a larger wave. *Destructive interference* occurs when the waves cancel one another out.

Something even more interesting occurs when two sounds with slightly different frequencies interfere. The combined sound rises and falls in amplitude. We hear this as a beat, a periodic change in the sound's loudness. Because the sounds do not share exactly the same frequency, destructive interference cannot completely cancel them. All it can do is change the loudness of the sound. The beats become longer and slower as the difference between the two frequencies decreases.

It's tempting to think that when waves come together the result will be bigger waves. But when waves from different sources exist at the same time in the same substance, what happens will depend on how their vibrations match up.

Light is a different story. Visible light is a mixture of frequencies. But light waves from two different sources cannot interfere with one another (except for lasers). This is why we don't notice interference patterns between streetlights or floor lamps. The only way to make light interfere is to split up the light from one source into separate beams. This is what happens when you see colors swirling over a soap bubble or an oil slick on water.

When light strikes a thin, transparent substance such as a soap bubble, it can be reflected at least twice—once by the outer surface and once by the inner surface of the bubble. Often the light is reflected many times. Whether light goes though constructive or destructive interference depends on the thickness of the film and the wavelength of the light. At any given point on the soap film, some wavelengths of light show constructive interference and other wavelengths show destructive interference. Because white light is made up of wavelengths of all colors, taking away some wavelengths and increasing others results in a colorful display.

The pearly, shimmering colors of mother-of-pearl also result from interference. Mother-of-pearl comes from the shells of oysters and abalones, which secrete thin layers of material on the inside surfaces of their shells.

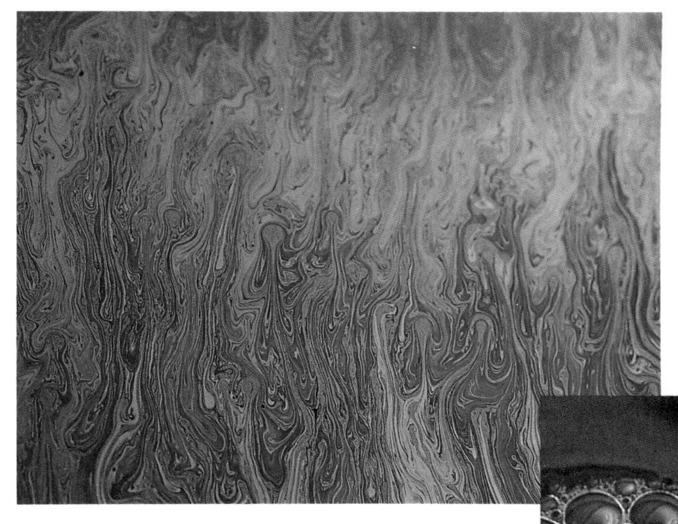

The beautiful, swirling colors on a soap bubble are caused by interference. If you look at a bubble close up, you can see the many colors of light.

Except for lasers, light waves from different sources cannot interfere with one another.

Mother-of-pearl is the shiny, colorful lining of an abalone or oyster shell. Interference is the reason for its changing colors.

POLARIZATION

The light given off by the sun and other incandescent sources—such as a candle flame—is always unpolarized. Its vibrations have no preferred direction and go every which way.

Bees use partly polarized light from the sun as a compass to help them find their way around.

Our eyes cannot see the difference between polarized and unpolarized light.

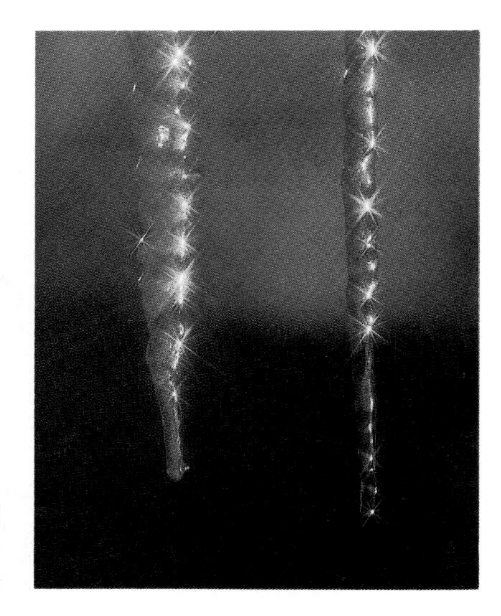

Many minerals—including snow and ice—can polarize light.

Have you ever wondered what polarized sunglasses are? The answer to this question has to do with the way in which light waves travel and their vibrations from side to side. You might imagine light waves traveling along side by side and all vibrating in the same direction. This is not the case. The truth is that light waves vibrate in many different directions perpendicular to their direction of travel. This means that if you could actually see individual light waves coming directly toward you from the sun, you would notice that some of them would be vibrating up and down, others would be vibrating side to side, and still others would be vibrating at all different angles. When light passes through certain kinds of matter, the waves can be made to vibrate all in the same direction. We say that the waves become *polarized* (PO-luh-reyezd).

Bees and some animals can see the difference between polarized and unpolarized light. The light from the sun becomes partly polarized when it is scattered by gas molecules in the atmosphere. Bees can use the polarized light to figure out the location of the sun even on cloudy days.

Polarization was first discovered in minerals. Calcite and quartz can split one light beam into two polarized beams. Other substances, such as the mineral tourmaline and quinine crystals, can absorb light vibrating in one direction much more strongly than light vibrating at right angles.

So how does polarization help us? The most common use for polarized light can be found in sunglasses. While sunlight isn't polarized, it can become partially polarized after reflecting off certain surfaces. These reflections can create a harsh glare that makes you squint. Polarizing sunglasses reduce the reflections by absorbing the polarized light. At the same time, they allow normal sunlight through.

With the help of polarized sunglasses, you can see examples of double refraction in frost. On a clear day, look through a frosted window while you rotate the sunglasses in front of your eyes. Polarized light from the sky passes through the tiny ice crystals and is bent by an amount that depends on the light's wavelength and direction of vibration. The frost will seem to be painted with faint splashes of color. If you put a few layers of cellophane tape on a clear surface, you will see the same effect.

Photographers also use polarizing filters. Light from the clear sky can be partly polarized. Photographers use a polarizer to reduce the polarized light from a scene. This makes the sky look darker in the photograph, but it does not darken the clouds by the same amount, so they stand out more.

And what about sound waves? Because sound waves in air are longitudinal, rather than transverse, there is only one possibility for vibration. This means that sound waves do not have the property of polarization.

You can see the effect of polarization on a hot highway in the "puddles" of light that seem to appear in the distance. The puddles are actually polarized reflections of the sun's light.

When you wear polarized sunglasses, you can still see objects around you, but the glare from the reflections is dimmed.

If you look at frost on a windowpane through polarized sunglasses, you can see the effects of double refraction.

LASER LIGHT

One of the most common lasers uses a combination of helium and neon gas—it makes the red light of the supermarket scanner.

Inside a laser tube, electrons are excited by an electrical current or a flash from a bright light.

"Laser" stands for "light amplification by stimulated emission of radiation."

Lasers have many uses in industry. They are very precise tools for cutting, welding, and many other operations.

Lasers produce a beam of light that is compact and intense, scientifically interesting, and very useful. In fact, you've probably seen a laser in action—it's the web of red light that scans bar codes at the supermarket checkout. Lasers are also used in compact disc players, so you may own one without knowing it.

What makes laser light so special? Unlike light from other sources, the light produced by lasers has only one very pure wavelength. Many gases and even solids and liquids can be used in producing laser light.

The light waves from lasers move together in a way that light from ordinary sources cannot. The sun or an incandescent lamp, for example, does not give off light as a steady stream of waves but rather in random bursts. The unique property of lasers is that their light waves move along in step—the light is said to be *coherent* (ko-HEER-uhnt).

At the heart of every laser is a tube containing some material whose atoms must first become excited. That is, the atoms must be placed in a state where their electrons are in high orbits and ready to give off photons by falling back to lower levels. Once in a high-energy state, the electrons would normally just give off their light and return to the lowest orbit. But under the right conditions, the presence of light can force the electrons to do this much sooner than normal. Light is amplified, or made stronger, by this process.

The material in the tube gives off one wavelength of light. Some of the waves strike a mirror at the end of the tube, which reflects them back into the material. As they travel along the tube, the reflected waves stimulate more atoms to emit radiation. A beam of coherent, amplified light builds up in the tube. When the beam reaches the front of the tube, a special kind of mirror lets some waves escape as an intense beam of light. Those that don't escape bounce back through the tube, exciting more atoms.

Because light vibrates so much faster than radio waves, lasers can be used to carry great amounts of information. Telephone and cable television companies pipe laser light through special fibers that trap the light inside by internal reflection. A 144-fiber cable can carry 40,000 telephone conversations at once!

Lasers have found uses in construction and surveying because they make a perfectly straight line. And since the beams travel at the speed of light, which is constant, they can be used to measure distance. Astronomers have used such "laser ranging" to determine the moon's distance from Earth. In manufacturing, laser beams can melt, weld, cut, or drill.

Even more remarkable, lasers can record three-dimensional images called *holograms.* A laser beam can be split into two halves by a mirror. These two halves are used to record a hologram of an object on photographic film.

Lasers and fiber optics can carry much more information than electrical currents in wires.

You've probably seen a special kind of picture called a hologram.

Above: Today you can find holograms on credit cards, jewelry, and magazine covers. Right: Lasers are used for storing and retrieving information on compact disks.

LIGHT AND THE UNIVERSE

are closely connected. Using light, scientists are working to solve important questions about the history of the universe and how it evolved. Because of our understanding of light, we now know what makes up the stars, galaxies, and clouds of dust drifting between the galaxies.

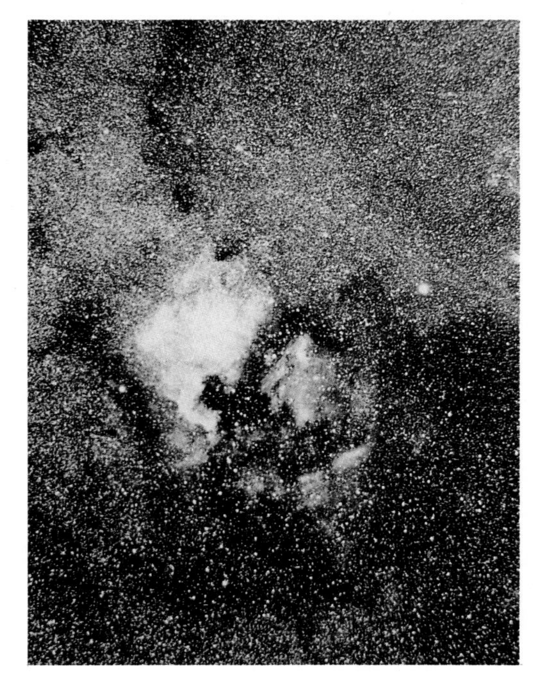

The study of light also tells us that our universe is expanding. And it could tell us how and when our sun and the stars came into existence. Light from the very edges of the visible universe began its journey to Earth millions of years ago and it carries information about events in that distant past. With telescopes and satellites, we use light to help us probe the secrets of the universe.

LIGHT AND ENERGY

Visible photons arrive from the sun in great numbers, bringing in vast amounts of energy.

The study of light is a very powerful tool indeed.

Photons can hit a specific molecule on a piece of photographic film to produce an image. This is something particles can do.

A spectrum, like this one, helps scientists study the colors of light coming from a distant star. This information helps them determine what makes up a star millions of miles away.

Light is one of the purest forms of energy that we encounter in our world. Photons are the basic bits of light energy. The energy of a photon of visible light is very tiny. On the other hand, a single gamma ray, which is a high-frequency photon, is quite energetic and can be detected easily. Gamma rays, like light, are part of the electromagnetic spectrum.

Photons are hard to define. In some ways they behave like waves, but in other ways they behave like particles. Photons give rise to interference, just as waves do. They also have wavelength and frequency measurements, just like waves. In certain experiments, however, scientists have found that photons can act as particles. For example, they can scatter or make a particle of mass move by bumping into it. The important things to remember about photons is that they have no mass and that they always travel at the speed of light.

A photon is emitted when changes take place within the atoms of a substance. The color of a photon depends on the substance that emitted it. For example, we know that sodium puts out a specific shade of yellow photons. Scientists often use light to study chemicals. The colors of light emitted by a substance can show us exactly what the substance is made of. That's how we know what the sun, the stars, and the intergalactic gases are made of. Scientists have studied the light these objects emit.

The energy in light also can be transformed into other forms. If you've ever walked on an asphalt driveway in bare feet on a sunny day, you know how hot the asphalt gets. You've also probably heard someone saying that you could fry an egg on the hood of a black car parked in the hot sun. This is partly because visible light from the sun is totally absorbed when it encounters a black surface. The visible solar light energy is transformed to heat energy. Infrared rays from the sun also add heat.

Solar light energy cannot normally be stored except indirectly. Water in a lake heats up over the summer as it absorbs solar energy. When winter arrives, the warm lake makes the weather milder near the shore. The sun's energy is stored in the water as heat, and then the heat is slowly lost over the winter. Solar cells convert solar energy into electrical energy that is stored in batteries. In photosynthesis, the energy of light is stored in the chemical bonds of the plant's food to be used later as energy. In all these cases, the light itself vanishes, transformed into a new state.

Light's energy can be transformed and used in many ways.

A tree captures the sun's energy and transforms it into the chemical energy the tree needs to survive and grow. This giant sequoia has been collecting and using solar energy for centuries.

Solar cells can collect energy from the sun and convert it into electrical energy. The electrical energy can be stored in batteries.

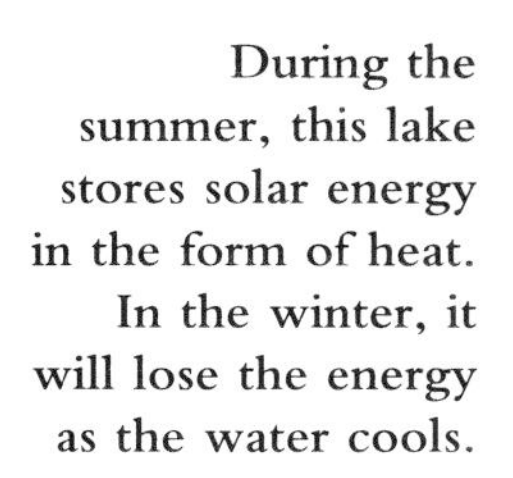

During the summer, this lake stores solar energy in the form of heat. In the winter, it will lose the energy as the water cools.

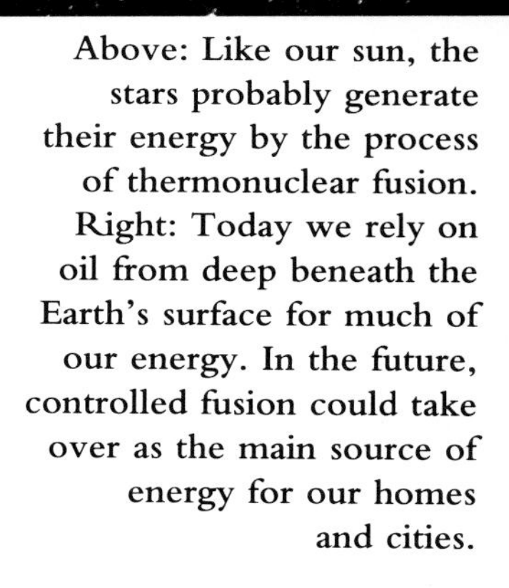

Once fusion has begun, the process keeps itself going.

Above: Like our sun, the stars probably generate their energy by the process of thermonuclear fusion. Right: Today we rely on oil from deep beneath the Earth's surface for much of our energy. In the future, controlled fusion could take over as the main source of energy for our homes and cities.

The sun is the source of our light on Earth. But how is the sun's energy generated? Scientists believe that a process called *thermonuclear fusion* (ther-mo-NOO-kleer FYOO-zhuhn) occurs deep inside the hot sun, where the temperature is about 27 million degrees Fahrenheit. Under these extremely high temperature conditions, the nuclei of hydrogen atoms are smashed together with great force. They stick together to form the nuclei of helium atoms. This process releases energy and maintains the intense high temperature. Some of the sun's matter is converted to energy.

If it takes high temperature to continue the process, how does it start? Astronomers tell us that stars, such as our sun, originally evolved from large clouds of hydrogen gas drifting in space. These clouds got heavier, and the force of gravity caused them to start becoming smaller. This process releases gravitational energy in the form of heat. At some stage in the process, the temperature became high enough to begin fusion.

The goal of much energy research during the past 50 years has been to make a practical controlled fusion process on Earth. If scientists succeed with this goal, we will have

a pollution-free source of energy for our homes, cities, and transportation. Recent experiments have shown that we are getting very close to achieving this goal.

Nuclear fusion can be very destructive if it is not controlled. The hydrogen bomb is an uncontrolled release of vast amounts of nuclear energy.

LIGHT AND THE DOPPLER EFFECT

One important feature of both sound and light is the effect motion has on the waves. This is called the *Doppler effect.*

You can hear the Doppler effect as a drop in pitch as a car speeds past. As the car comes toward you, the waves press up closer to one another. This results in a shorter wavelength and a higher frequency. The opposite happens when the car has passed by: The waves get more stretched out.

The same thing happens with light. Approaching objects, such as stars, cause shorter wavelengths, so the light they emit goes farther into the blue part of the light spectrum. This is called the *blue shift.* Objects moving away cause a shift to longer wavelengths, or the *red shift.*

Doppler radar uses this principle to look at moving objects. When radar microwaves reflect off the targets, they are Doppler-shifted in frequency. Wave detectors can determine how fast the target is moving.

Scientists have used the Doppler effect to find that light from all galaxies is red-shifted. This means that all the galaxies and stars are moving away from us. Scientists see this as evidence that the universe is expanding. By figuring out how fast the universe is expanding, we can backtrack to estimate the age of the universe.

You have probably experienced the Doppler effect. You hear it in the quick change in pitch of a train horn or police siren as it rushes past you.

In order for us to see the Doppler effect on light with the naked eye, an object must be traveling at least one percent of the speed of light. Nothing humans have invented can move that fast!

Baseball pitchers often have their pitches clocked by hand-held Doppler radar "guns" behind the plate.

Doppler radar can be used to track the progress of a tornado.

GLOSSARY

Acoustics (uh-KOO-stiks): The science that studies sound.

Amplitude (AM-pluh-tood): The height of a wave's crest above its center.

Chlorophyll (KLOR-uh-fil): A pigment in green plants that enables them to use the energy in sunlight to produce food.

Decibel (DE-suh-bel): A scale used to measure the loudness of sounds.

Doppler effect: A change in the frequency of light or sound waves that is observed when the wave source moves.

Frequency (FREE-kwuhn-see): The measure of the number of wave cycles that pass by a certain point in one second.

Incandescence (in-kan-DES-uhns): The emission of visible light by a body, caused by its high temperature.

Interference: The effects of two waves combining.

Laser: A device that produces a coherent beam of light by exciting atoms and causing them to emit energy in regular waves, rather than randomly, as in ordinary light.

Longitudinal (lawn-juh-TOOD-nuhl) wave: A wave that moves by compressing, then stretching out the substance through which it travels. Sound waves are longitudinal when they move through air.

Luminescence (loo-muh-NE-suhns): The emission of light that is not caused by incandescence and that occurs at low temperatures.

Photon: A quantity of light energy emitted by an atom when certain changes take place inside the atom.

Photosynthesis (fo-to-SIN-thuh-suhs): The way in which green plants form food from carbon dioxide and water using sunlight as the source of energy.

Polarization (po-luh-ruh-ZAY-shuhn): The effect of causing light waves to vibrate in the same direction. Polarized sunlight can cause a harsh glare.

Radiation (ray-dee-AY-shuhn): Energy released by an object in the form of waves and particles. Light is a form of radiation that we can see.

Refraction (ri-FRAK-shuhn): The bending of a wave as it moves from one substance to another, such as from air to water or from warm air to cold air. Refraction of light can create rainbows.

Reflection: The return of light or sound after it strikes a surface.

Resonance (REZ-uhn-uhns): A larger-than-normal vibration that takes place when a vibrating object, such as a musical instrument, receives pulses of energy at its natural frequency.

Spectrum: A grouping of light or sound waves by frequency or wavelength. The electromagnetic spectrum is the colored band of visible light wavelengths (red, orange, yellow, green, blue, indigo, and violet) produced by a prism, together with invisible extensions (such as radio, infrared, ultraviolet, and X ray). The sound spectrum humans can hear is the range of frequencies between about 20 Hz and 20,000 Hz.

Transverse (trans-VERS) wave: A wave that moves at right angles to its direction of travel. Light waves are transverse waves.

Wavelength: A measure of the distance a wave travels during a single cycle.

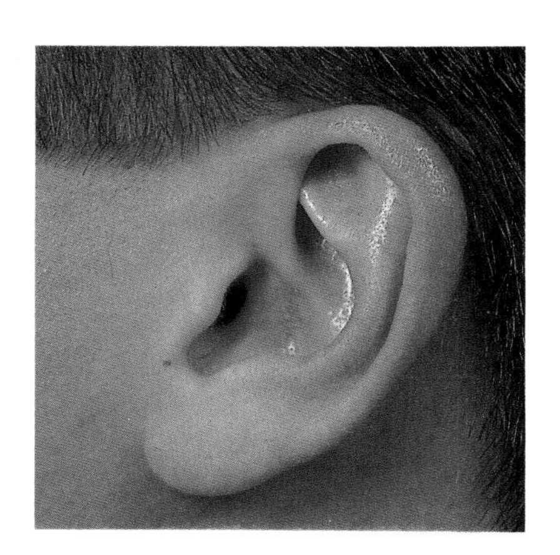